Helping the Community

# What Do GARBAGE COLLECTORS Do?

Richard Alexander

PowerKiDS press
New York

Published in 2016 by The Rosen Publishing Group, Inc.
29 East 21st Street, New York, NY 10010



First Edition

Editor: Katie Kawa
Book Design: Katelyn Heinle

Photo Credits: Cover (garbage collector), pp. 1, 13 Don Mason/Blend Images/Getty Images; cover (hands) bymandesigns/Shutterstock.com; back cover Zffoto/Shutterstock.com; p. 5 360b/Shutterstock.com; p. 6 Dmitry Kalinovsky/Shutterstock.com; p. 9 Johnny Habell/Shutterstock.com; p. 10 © iStockphoto.com/kozmoat98; pp. 14, 17, 18 © iStockphoto.com/PeopleImages; pp. 21, 24 (landfill) yevgeniy11/Shutterstock.com; p. 22 © iStockphoto.com/whitemay; p. 24 (garbage) Nataliya Hora/Shutterstock.com.

Library of Congress Cataloging-in-Publication Data

Alexander, Richard, author.
What do garbage collectors do? / Richard Alexander.
pages cm. — (Helping the community)
Includes bibliographical references and index.
ISBN 978-1-4994-0631-3 (pbk.)
ISBN 978-1-4994-0632-0 (6 pack)
ISBN 978-1-4994-0634-4 (library binding)
1. Sanitation workers—Juvenile literature. 2. Refuse and refuse disposal—Juvenile literature. I. Title.
HD8039.S257A44 2016
628.4'42—dc23

2015000445

Manufactured in the United States of America

CPSIA Compliance Information: Batch #WS15PK: For Further Information contact Rosen Publishing, New York, New York at 1-800-237-9932

# CONTENTS

**Garbage** collectors work very hard.

They keep our towns clean.

Garbage collectors pick up garbage. People leave their garbage in cans near the street.

Garbage collectors put the garbage into a big truck.

One worker drives the truck.

Other workers pick up the garbage.

The workers sometimes stand on the back of the truck as it moves.

The workers wear bright colors so drivers can see them.

They take the garbage to a **landfill**. This is a place with a lot of garbage.

Garbage collectors have an important job to do!

# WORDS TO KNOW

garbage

landfill

# INDEX



# WEBSITES

Due to the changing nature of Internet links, PowerKids Press has developed an online list of websites related to the subject of this book. This site is updated regularly. Please use this link to access the list: www.powerkidslinks.com/htc/gcol